AI AWARENESS SERIES

AI in Banking
and Finance

Andrea V.M. Greaves

Contents

Introduction

Artificial intelligence is no longer a future concept in banking and finance—it's here, redefining how financial institutions operate, interact with customers, and make decisions. From real-time fraud prevention to hyper-personalized financial services, AI is reshaping the industry at every level.

This book, AI in Banking and Finance, explores the growing influence of AI across the financial ecosystem. It provides a practical and strategic view of how data-driven intelligence is enhancing operational efficiency, enabling smarter risk management, and powering innovation in product development and customer engagement.

We begin with a broad overview of how AI is being applied across financial services, followed by a deep dive into the critical importance of high-quality data foundations. Subsequent chapters explore key AI applications such as fraud detection, credit risk scoring, stress testing, market trend analysis, and portfolio management. We also examine how AI is supporting sustainable investing, driving behavioral insights, and powering virtual financial advisors.

In parallel, the book addresses vital issues of fairness, bias, and responsible AI use, recognizing the ethical and regulatory challenges that come with increased automation. We also look at how AI is augmenting human roles, transforming organizational structures, and guiding digital transformation strategies.

Chapter 11: Market Intelligence and Agribusiness Innovation

This book is designed for professionals across the financial industry—whether you're in risk, compliance, strategy, IT, customer service, or investment. It's also a valuable resource for students, policymakers, and technologists who want to understand how AI is practically deployed in one of the most data-intensive and regulated sectors of the economy.

The goal is not just to showcase what AI can do, but to help you think critically about how it should be used—responsibly, effectively, and strategically.

Let's begin.

Chapter 1: Overview of AI in Financial Services

AI is reshaping the financial sector in several key ways. It allows faster decision-making by rapidly analyzing financial data, which leads to quicker and more accurate insights. In risk management, AI helps detect patterns and predict potential financial threats. Customer service is also enhanced through AI-powered tools that offer personalized support. And importantly, adopting AI drives efficiency and innovation, giving financial institutions a strong competitive advantage.

Machine learning is about teaching algorithms to learn from data. These models analyze data, find patterns, and then use those patterns to make predictions. In finance, machine learning powers AI applications that improve decision-making, enhance risk management, and automate routine tasks, leading to more efficient financial operations.

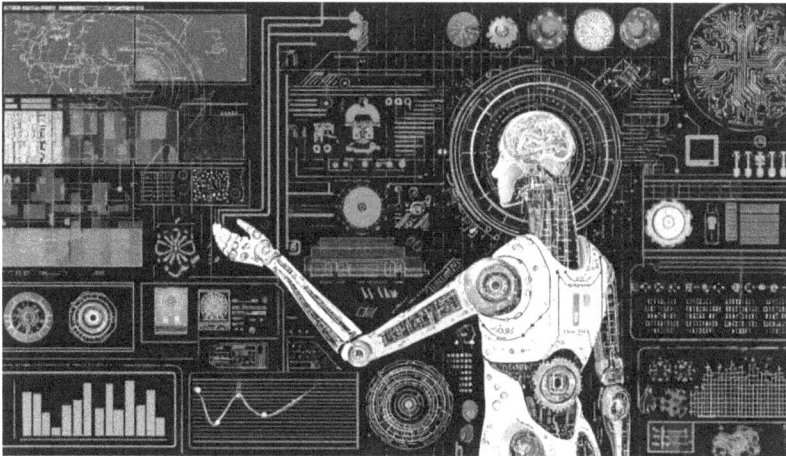

Deep learning builds on machine learning by using neural networks to understand complex patterns in large datasets. This technology supports tasks like advanced image recognition and voice analysis. For the finance industry, deep learning is driving innovation—helping institutions analyze market trends, customer behavior, and financial risks with greater accuracy.

Natural Language Processing, or NLP, allows computers to understand human language. In finance, this means quickly analyzing documents to extract key information, using sentiment analysis to gauge market sentiment, and powering chatbots that improve customer interaction. NLP is a vital tool for making sense of large volumes of text data efficiently.

Structured financial data—like statements, transactions, and market data—is highly organized, making it easier to analyze. AI models use this data for forecasting trends, assessing risks, and detecting anomalies. These applications help financial institutions make better-informed decisions and protect against potential threats.

Unstructured data includes sources like news articles, reports, and customer communications, which don't follow a set format. NLP techniques are essential here, helping extract insights and identify patterns from these text-heavy sources. These insights can influence investment decisions and help monitor compliance risks.

Before AI models can work effectively, financial data needs to be cleaned and prepared. Data cleaning removes errors, while feature engineering creates meaningful variables that make AI models more accurate. Given the variety of financial data formats, preprocessing is a crucial step that directly impacts model performance.

Supervised learning is used when we have labeled data, such as historical transactions for fraud detection or credit data for scoring purposes. These models learn patterns from past data to predict future outcomes, helping institutions identify fraud or assess credit risk accurately.

Unsupervised learning works with unlabeled data. It's commonly used for anomaly detection—spotting unusual activities that may indicate fraud—and for customer segmentation, where hidden patterns in behavior are used to tailor marketing and service strategies. This helps institutions better understand and serve their clients.

Choosing between supervised and unsupervised learning depends on factors like the availability and quality of data, the complexity of the task, and the alignment with business goals. It's also important to understand each method's limitations and risks to ensure successful and responsible AI deployment.

Natural Language Processing can automate many compliance tasks. It helps extract relevant data from financial documents, allowing faster, more accurate financial analysis. This automation supports timely decision-making and improves risk assessment by analyzing financial disclosures and news efficiently.

NLP also enhances transparency and auditability. It tracks changes in documents automatically, making it easier to maintain audit trails. Automated summarization helps with quick reviews during audits, while better traceability ensures that compliance actions are well-documented and easy to verify.

Large foundation models, including large language models or LLMs, have become central to AI in finance. These models are trained on vast datasets and can be fine-tuned for finance-specific tasks. They form the backbone of many advanced AI applications in the industry.

By fine-tuning LLMs with financial data, we can create powerful tools for specific financial tasks. These models can automate detailed report generation, offer personalized customer support, and enhance risk analysis. This customization makes AI tools more relevant and effective in financial contexts.

Looking ahead, the use of large language models in finance will only grow. However, it's critical to emphasize ethical considerations, especially around data privacy. Ensuring transparency and addressing biases in AI models will be key to building trust and encouraging responsible adoption across the industry.

To wrap up, we've explored how AI is transforming the financial sector—from foundational concepts to practical applications in data analysis, compliance, and customer service. We've also looked at emerging trends like large language models and discussed important ethical considerations. Understanding these concepts is crucial for finance professionals aiming to leverage AI effectively in their work.

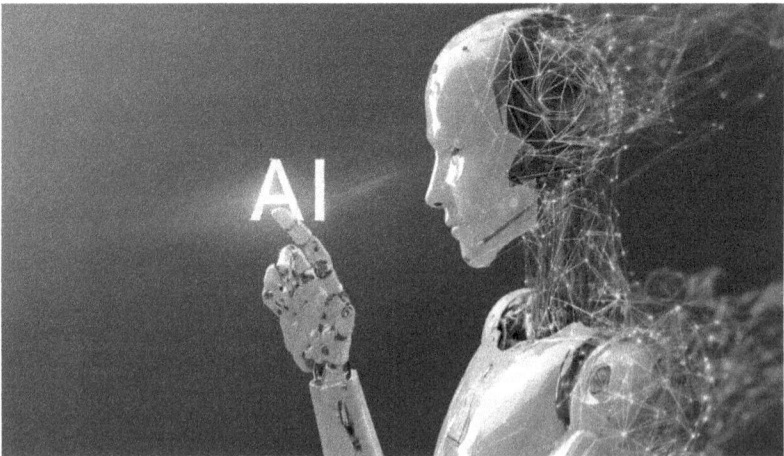

Chapter 2: Data Foundations for Financial AI

Traditional data warehouses are strong when it comes to storing and managing structured data — they offer reliable governance and powerful analytics tools for business decisions. However, they aren't built for unstructured data like text or multimedia. Plus, when it comes to scaling for AI workloads, traditional warehouses often fall short in performance and flexibility.

To address those limitations, organizations are moving towards data lakes and federated systems. Data lakes provide scalable storage for diverse and large datasets, making them ideal for AI applications. Federated systems allow for secure data sharing without the need to centralize sensitive information, which helps protect privacy. When combined, these approaches offer flexible AI model development while keeping sensitive data secure.

Federated data approaches bring unique benefits for AI applications. They allow AI models to train on distributed datasets without compromising privacy, which is crucial in regulated industries. Since data stays on local servers, this also reduces regulatory risk related to data transfers. Additionally, federated systems make it easier for organizations to collaborate, sharing insights without having to exchange raw data.

Data lakes serve as powerful storage platforms, handling massive volumes of both structured and unstructured data. They act as a foundation for advanced analytics and AI, and they scale cost-effectively as data needs grow. This flexibility makes them essential for modern AI-driven organizations.

Synthetic data plays a critical role in AI development. It mimics the patterns of real data without exposing sensitive information, making it a valuable tool for privacy compliance. Beyond that, synthetic data enables the development and testing of AI models in a controlled, ethical manner.

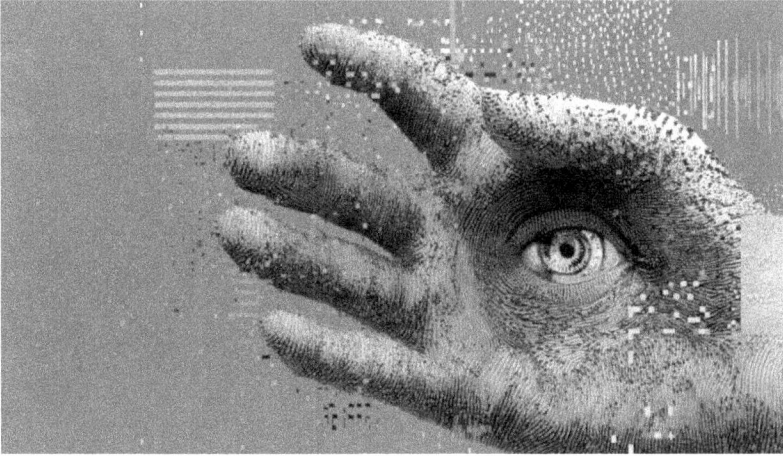

Secure sandboxes are crucial for AI experimentation. They create isolated environments that keep AI tests separate from live systems, reducing risk. These sandboxes tightly control data access, helping prevent unauthorized use or data leaks. This ensures that AI development can proceed safely while maintaining regulatory compliance.

Let's look at some key regulatory frameworks shaping AI systems. GDPR enforces strict data privacy rules, ensuring individuals' rights are protected. DORA focuses on enhancing the digital resilience of financial institutions, especially against cyber threats. Basel III strengthens banking capital requirements and promotes financial stability — all crucial when integrating AI into financial services.

To build AI systems that comply with these regulations, organizations need to embed compliance into their infrastructure design. This means ensuring data is handled according to privacy laws, implementing strict access controls, keeping detailed audit trails for accountability, and applying proactive risk management strategies.

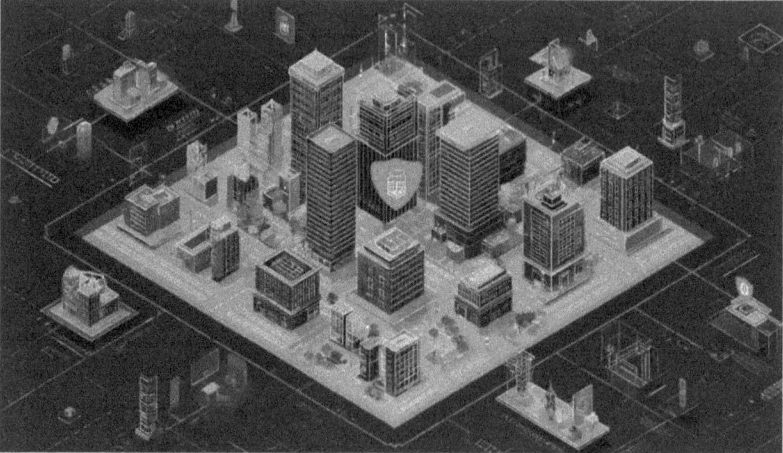

Compliance isn't a one-time task. It requires ongoing monitoring to catch issues early and adapt to new regulations. Regular audits help verify compliance, and staff training fosters a culture of regulatory awareness. Finally, flexible policies allow organizations to respond quickly to changes in the regulatory landscape and technological advancements.

Metadata is vital throughout the AI lifecycle. It documents where data comes from, tracks any changes it undergoes, and supports governance by providing insights into how data is used. This traceability ensures that AI projects are transparent and that data integrity is maintained.

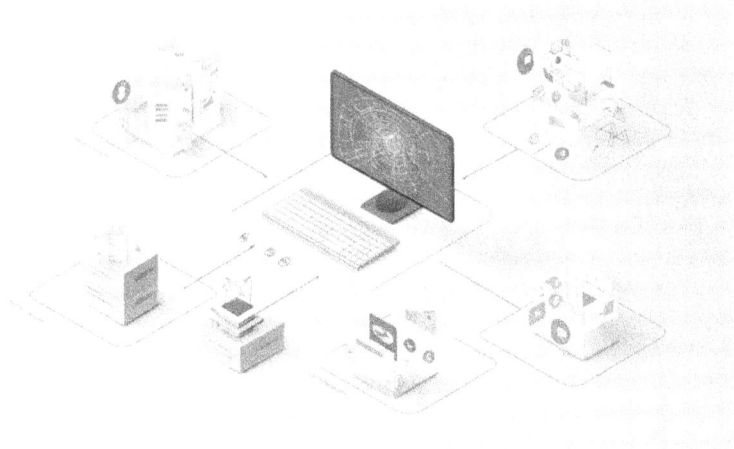

Data lineage takes metadata a step further by tracking how data moves and transforms across systems. This visibility helps ensure transparency in AI processes, allowing teams to identify the source of errors or biases and maintain trust in AI outcomes.

Metadata also plays a key role in making AI systems explainable and auditable. By providing context on AI outputs, metadata helps users understand and trust AI decisions. It also supports compliance audits by documenting the processes behind AI models, ensuring that ethical and legal standards are met.

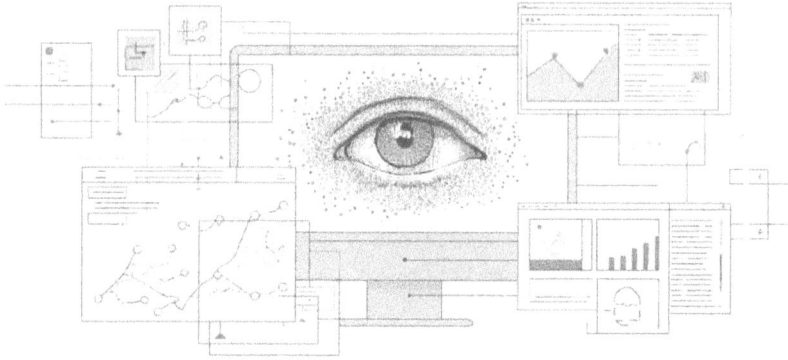

APIs are critical for seamless data integration. They enable AI systems to connect with various data sources, applications, and platforms — making interoperability and data flow much smoother across different environments.

Open banking is a prime example of data access and innovation in action. It allows for secure sharing of financial data with authorized third parties, fueling AI-powered financial services. This openness drives innovation, enhances financial decision-making, and ultimately provides consumers with more choices and better services.

Achieving interoperability is about ensuring AI systems can work together, regardless of platform or technology. This not only supports scalable deployments but also allows flexible integration into different environments. Interoperability is key to unlocking the full potential of AI in a connected world.

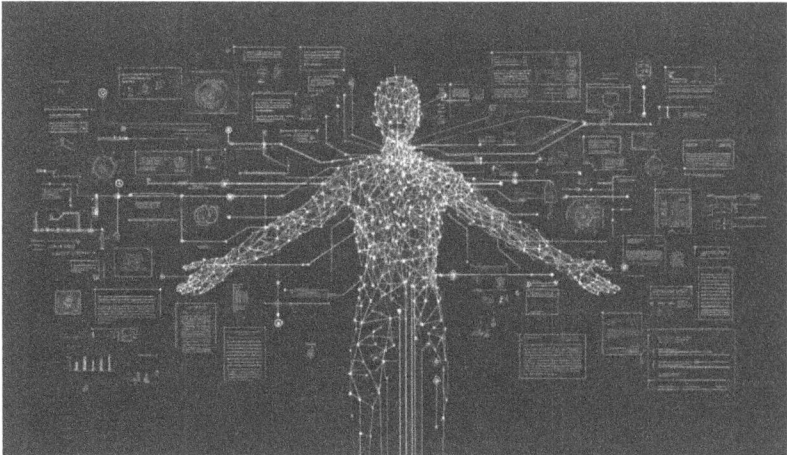

To wrap up, strong data infrastructure and regulatory readiness are essential for the successful deployment of AI systems. By leveraging modern data strategies, ensuring compliance, managing metadata effectively, and promoting interoperability, organizations can build AI solutions that are both powerful and responsible.

Chapter 3: AI for Real-Time Fraud Prevention

Let's begin with anomaly detection in payment and transfer systems. This involves techniques like threshold-based rules, which flag transactions that exceed certain predefined limits. These are useful for quick checks but are quite basic. Clustering techniques group similar transactions together, helping us spot outliers that don't fit the usual patterns. Predictive modeling, on the other hand, uses past transaction data to forecast what normal activity looks like—so when something deviates, it's quickly flagged.

Machine learning plays a critical role in real-time anomaly detection. It can analyze large volumes of transaction data on the fly, spotting irregularities with high efficiency. What's more, these models continuously improve as they learn from fresh data, which sharpens both their accuracy and their speed over time.

Let's look at some real-world impacts of anomaly detection. In finance, these systems help institutions quickly spot suspicious patterns and stop fraud before it happens. They can significantly reduce financial losses by intercepting fraudulent transactions early. Overall, anomaly detection strengthens security measures, making financial operations safer and more resilient.

Moving on to transaction pattern analysis and behavioral biometrics. First, transaction pattern recognition helps us understand typical user behavior—like spending habits and regular transaction types. When we see deviations from these patterns, it can serve as an early warning sign of fraud. So, this kind of analysis is a key part of effective fraud prevention strategies.

Behavioral biometrics takes this a step further by focusing on how users interact with systems. It uses patterns like typing rhythm or mouse movements for verifying identity. Because it allows for continuous authentication in the background, it enhances security beyond just the login stage. Importantly, this method balances stronger security with a smooth user experience.

Transaction pattern analysis and behavioral biometrics don't work in isolation. They are integrated into fraud detection workflows to complement other tools and systems. This integration ensures a layered security approach—making it harder for fraudulent activities to slip through the cracks.

Let's talk about AI-enhanced card security and geolocation triggers. AI is used for real-time monitoring of card transactions, scanning constantly for suspicious activity. It employs advanced algorithms that can detect unusual patterns in card usage. When something suspicious is flagged, the system can automatically alert the user or even block the transaction to prevent fraud.

Geolocation data also plays a crucial role in fraud prevention. By verifying that the location of a transaction matches where the user typically operates, the system can validate legitimate transactions. On the flip side, if a transaction occurs in an unexpected location, the system can flag it for investigation—or stop it altogether.

A key challenge here is balancing strong security with a positive user experience. AI-driven security must be robust enough to block fraud but also smooth enough to avoid frustrating genuine users. This balancing act is critical—ensuring both protection and customer satisfaction.

Now let's explore graph-based fraud detection and network mapping. Graph theory helps us analyze relationships between different entities, like users and transactions. By studying these connections, we can detect suspicious clusters of activity—often revealing hidden fraud rings. Identifying patterns in transaction links is a powerful way to expose coordinated fraud schemes.

Mapping transaction networks can help uncover fraud rings by showing how suspicious entities are connected. These visualizations give investigators a clear view of potential fraud networks, making it easier to take action.

However, graph-based detection comes with challenges. Handling massive data volumes and the complexity of graph algorithms can be demanding. Still, despite these hurdles, graph-based approaches have proven highly effective in real-world fraud prevention—especially for catching organized fraud.

Finally, let's talk about continual learning for adaptive fraud models. Continual learning allows AI systems to update their knowledge over time without starting from scratch. This helps models stay current and responsive as fraud tactics evolve.

Adaptive models are essential for dynamic fraud detection. They adjust and learn as new fraud patterns emerge, keeping detection capabilities sharp. By staying adaptable, these models are better equipped to handle changing tactics used by fraudsters.

There are clear benefits—and also some challenges—with continual learning. On the plus side, it improves how quickly and accurately models respond to new threats. However, models can suffer from catastrophic forgetting, where new learning overwrites previous knowledge. Maintaining quality data is critical to prevent this. Fortunately, with the right strategies, these limitations can be managed, ensuring continual learning remains a valuable tool in fraud prevention.

To conclude, AI is transforming real-time fraud prevention with advanced strategies like anomaly detection, pattern analysis, behavioral biometrics, graph-based detection, and continual learning. Together, these approaches create a robust, adaptive defense against evolving financial threats. As fraud tactics change, AI's role in safeguarding financial systems will only become more essential.

Chapter 4: AI-Powered Credit Risk Assessment

Let's begin by looking at how credit scoring is being modernized through machine learning models. These approaches move beyond traditional scoring by leveraging complex data patterns, offering more adaptive and accurate assessments of credit risk.

Traditionally, credit scoring relied on linear models using limited data. Machine learning models allow us to capture nonlinear patterns and interactions within data, leading to more accurate risk predictions. This shift enables better handling of diverse borrower profiles and dynamic financial behaviors.

Some key machine learning algorithms used in credit risk assessment include logistic regression, which handles binary outcomes like default risk; decision trees and random forests, which improve accuracy through ensemble methods; gradient boosting, which combines weak learners into strong predictors; and neural networks, which excel at capturing complex nonlinear relationships in credit data.

Machine learning credit scoring brings clear benefits—improved accuracy, adaptability to new data, and greater efficiency in decision-making. However, there are also limitations, such as dependence on high-quality data, the risk of overfitting, and challenges in interpreting more complex models.

Alternative credit scoring leverages non-traditional data sources like social media, mobile usage, and utility payments. This approach enriches credit profiles, especially for consumers with little or no conventional credit history, enhancing inclusivity and improving risk assessment accuracy.

By integrating social media, mobile, and utility data, we can build more comprehensive insights. Data fusion techniques enhance model robustness, feature engineering extracts meaningful predictors from diverse sources, and complex data pipelines ensure efficient processing of these varied data streams.

However, using alternative data introduces privacy and regulatory concerns. It's essential to address privacy proactively, navigate strict regulatory scrutiny, and ensure compliance with data protection laws. Transparency and adherence to regulations are key to maintaining consumer trust.

Explainable AI, or XAI, plays a critical role in credit decisions. Transparency allows consumers to understand why decisions are made, enables regulators to verify fairness, and supports ethical use of AI in financial services—ultimately ensuring trust and accountability.

Several techniques help explain AI model decisions. SHAP values clarify feature contributions, LIME explains predictions on a local level by simplifying complex models, and rule extraction transforms complex models into understandable decision rules. These tools make AI outputs more interpretable.

Providing clear explanations builds confidence among regulators and consumers alike. Transparent AI fosters trust, supports responsible lending practices, and ensures financial decisions are both ethical and effective.

Real-time data streams allow dynamic risk assessment. By continuously analyzing transactions and other inputs, models can instantly update risk profiles, enabling lenders to respond swiftly to changing financial conditions and emerging risks.

AI-driven portfolio management enhances risk oversight. These tools provide insights into overall portfolio exposure, help identify areas of concentrated risk, and allow continuous monitoring of exposure trends. This supports proactive portfolio adjustments and optimized lending strategies.

Predictive analytics plays a vital role in risk mitigation. By forecasting potential defaults and emerging credit risks, lenders can take early action to protect portfolio health, reduce losses, and manage risk more proactively.

Bias in AI credit models often stems from data imbalances, historical prejudices reflected in training data, and flawed feature selection. Identifying these biases is crucial to avoid perpetuating unfair or discriminatory lending practices.

To ensure fairness, techniques like fairness-aware algorithms, bias detection metrics, and post-processing adjustments are used. These approaches help correct biases either during or after model development, supporting more equitable credit decisions.

However, challenges remain. Balancing accuracy with fairness is a constant concern. Transparency in algorithms is vital for trust, and adherence to ethical regulations is critical for responsible AI deployment. Ongoing efforts are needed to meet both regulatory and societal expectations.

To conclude, AI is transforming credit risk assessment with powerful tools for prediction, decision-making, and fairness. But with this power comes the responsibility to use AI ethically—ensuring transparency, fairness, and compliance at every stage of the credit lifecycle.

Chapter 5: Stress Testing and Scenario Modelling

Stress testing and scenario modelling are critical tools for financial institutions. They help evaluate how an organization might perform under adverse conditions. These models test resilience against market shocks, economic downturns, or other stress events, which is essential for sound risk management and regulatory compliance.

Scenario modelling serves several key purposes. First, it involves creating hypothetical situations to evaluate potential risks and strategies. This helps institutions test their risk exposure under various future conditions. Ultimately, the goal is to support better decision-making by anticipating how the future might unfold and preparing for different outcomes.

Traditional modelling approaches come with several challenges. Incomplete data often leads to inaccurate predictions. Linear assumptions limit the ability to capture complex market behaviors.

And perhaps most critically, traditional models struggle to adapt to evolving and dynamic financial conditions, making them less effective in today's fast-paced environment.

Macroprudential risk focuses on threats that affect the entire financial system, not just individual institutions. Managing these risks is crucial for maintaining financial stability, preventing economic disruptions, and supporting market confidence.

AI brings powerful tools to systemic risk detection. It can detect anomalies that signal systemic threats. Network analysis maps relationships and uncovers interconnected risks. Predictive modelling uses AI algorithms on historical and real-time data to forecast emerging risks, providing early warnings that traditional methods may miss.

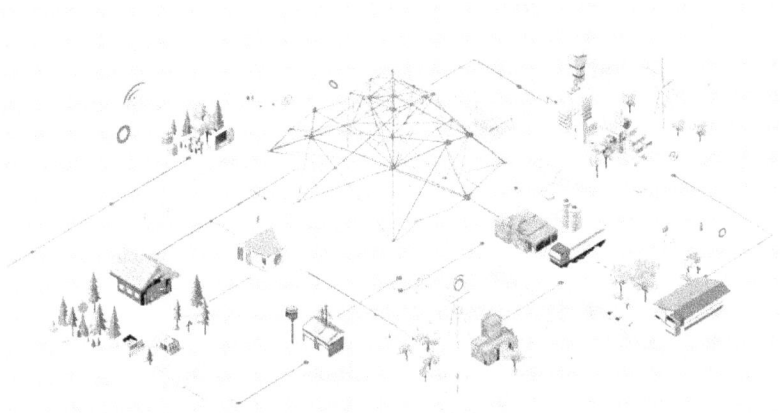

Financial institutions increasingly integrate AI into macroprudential modelling. AI enhances their ability to monitor systemic risks and strengthens regulatory responses. These advanced tools help regulators act earlier and more effectively to maintain stability in the financial system.

Generative AI models are designed to create new data by learning from existing datasets. Technologies like GANs and VAEs are especially useful for simulating complex financial data scenarios. These models allow us to simulate financial crises that go beyond historical patterns, providing valuable insights for risk preparedness.

Generative AI helps create hypothetical crisis scenarios, allowing institutions to consider risks outside historical experience. These AI-driven simulations include unprecedented stress events, enabling better anticipation of novel risks. This supports proactive risk management and improved resilience planning.

AI-generated crisis scenarios are vital tools for evaluating the resilience of financial institutions. By testing against these AI-generated events, firms can identify vulnerabilities and strengthen their risk management frameworks against a wider range of possible threats.

AI agents offer the ability to simulate markets in real time. They model complex financial environments and reflect rapid market changes, providing continuously updated stress scenarios. This real-time approach supports more effective and dynamic risk management strategies.

Adaptive stress testing methodologies use feedback loops to refine accuracy over time. Machine learning algorithms analyze patterns and predict failure modes, enhancing the predictive capabilities of stress tests. This adaptive approach improves decision-making under uncertain conditions.

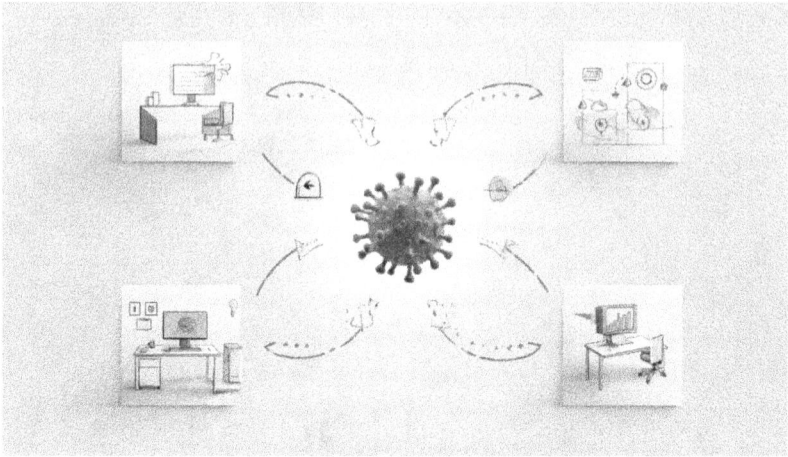

Dynamic stress testing enables continuous portfolio evaluation under various stress conditions. These methods help assess vulnerabilities, enhance portfolio resilience, and support informed risk management decisions. Real-time data allows for timely strategic adjustments, optimizing portfolio outcomes in the face of market volatility.

By integrating economic models with machine learning, we can capture complex economic relationships that traditional models miss. This approach adapts to new data insights dynamically and overcomes the limitations of relying solely on theory or empirical data.

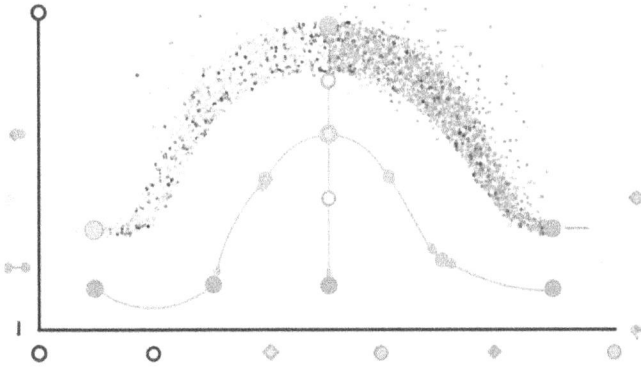

Hybrid models offer enhanced flexibility by combining different modelling approaches. They improve interpretability by blending understandable components with advanced algorithms. However, they can become complex and computationally demanding, which are important factors to consider when designing and implementing them.

Hybrid stress test frameworks enhance scenario analysis by integrating multiple risk factors and stress conditions. They improve risk quantification and are widely applied in finance sectors like banking, insurance, and investments to manage risks more effectively.

Climate risk presents significant challenges for financial markets and economic stability. Its complex, uncertain nature requires sophisticated modelling tools. Financial models play a key role in developing strategies to mitigate the economic risks posed by climate change.

AI significantly broadens the range of climate-related scenarios we can explore. It captures uncertainties more effectively than traditional models and can model complex climate system interactions that classical approaches may overlook, leading to better-informed risk assessments.

Regulations increasingly demand detailed climate risk disclosures. AI supports these requirements by enhancing scenario analysis capabilities, enabling institutions to provide more thorough and meaningful climate risk assessments in their reporting.

In conclusion, AI is transforming stress testing and scenario modelling in financial risk management. It enables more accurate risk detection, dynamic scenario generation, and comprehensive analysis of emerging threats. Whether in macroprudential risk, crisis simulation, portfolio management, hybrid modelling, or climate risk analysis, AI is an essential tool for navigating today's complex financial landscape.

Chapter 6: Predictive Analytics for Market Trends

Predictive analytics in financial markets relies on both historical and real-time data to forecast trends and behaviors. It involves a wide range of analytical techniques — from traditional statistics to cutting-edge machine learning. The ultimate goal is to deliver actionable insights that help traders and investors make better decisions.

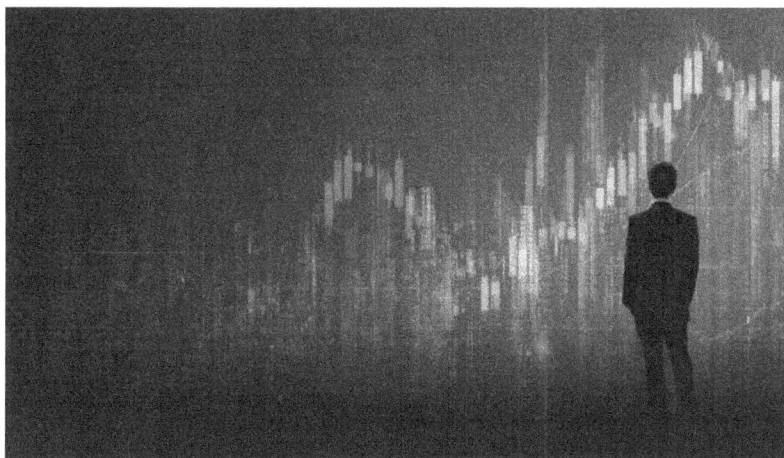

However, forecasting market movements comes with challenges. Market volatility leads to unpredictable price swings, demanding adaptive strategies. Data noise can obscure meaningful patterns, making it harder to identify true signals. Non-stationarity means the statistical properties of market data change over time, complicating model predictions. And the complex interdependencies between market factors require sophisticated modelling approaches.

Several emerging technologies are addressing these challenges. Deep learning models can identify intricate data patterns, boosting prediction accuracy. Large language models improve our understanding of market sentiment by analyzing economic news. And real-time signal processing allows traders to react quickly with sophisticated, data-driven strategies.

Traditional time series forecasting methods assume linear patterns, which limits their effectiveness in dynamic markets. Deep learning models, on the other hand, can capture nonlinear relationships and long-term dependencies, making them better suited for financial forecasting.

LSTM and GRU models are types of recurrent neural networks designed to process sequential data like financial time series. Their ability to remember long-term patterns makes them especially useful for predicting trends and price movements over time.

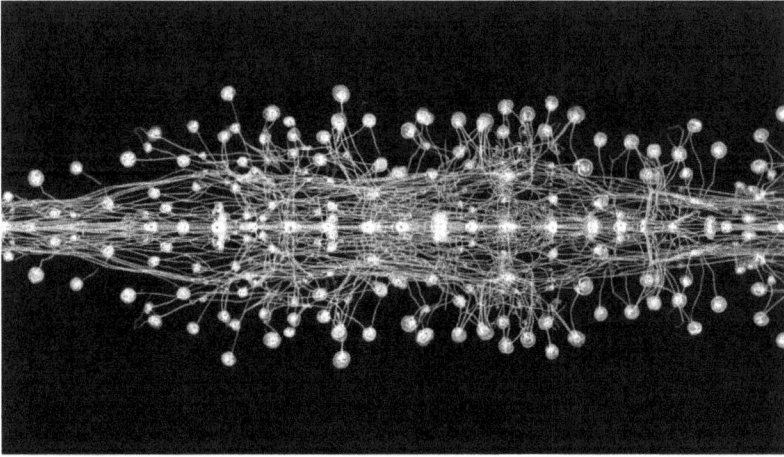

When evaluating deep neural forecasting models, we look at metrics like RMSE for accuracy and directional accuracy for trend prediction. But these models also have limitations — overfitting can reduce their ability to generalize, and their complexity can make them hard to interpret. Plus, they require significant computational resources, which can impact efficiency.

Large language models are powerful tools for summarizing economic news and extracting meaningful insights. They can process vast amounts of unstructured data, distilling relevant information that traders can act upon.

These models don't just summarize — they also extract actionable trading signals from economic news. This helps traders identify opportunities and risks hidden within complex narratives.

Integrating summarized news signals into quantitative trading models can make them more responsive to market events. By incorporating these insights, trading algorithms can react faster and predict market movements more effectively, enhancing their overall performance.

Financial markets generate diverse data — from news articles to social media sentiment and price feeds. Textual news provides context, social sentiment captures emotional trends, and quantitative price data offers critical technical insights. Understanding all three is key to a comprehensive market analysis.

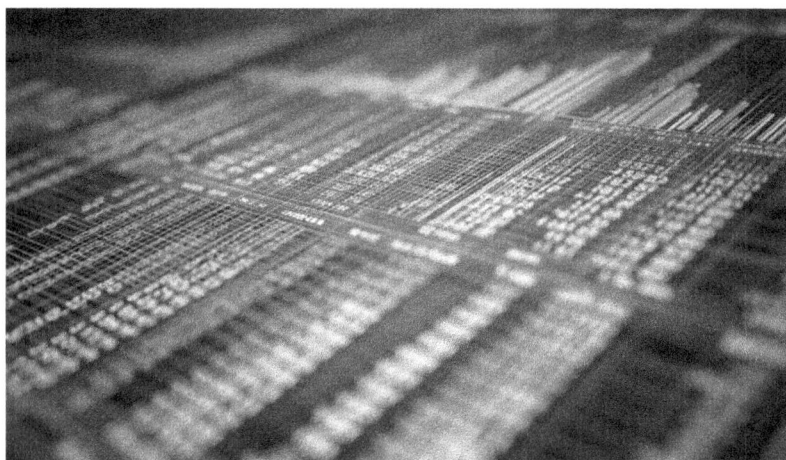

There are several ways to combine these multimodal data sources: Feature-level fusion integrates raw data features into a single model input. Decision-level fusion combines outputs from different models for a stronger forecast. And model-level fusion merges different forecasting models into a unified system for better analysis.

By fusing multimodal data, we enhance the context and depth of market insights. This leads to more accurate predictions and greater model robustness, especially in volatile or noisy market conditions.

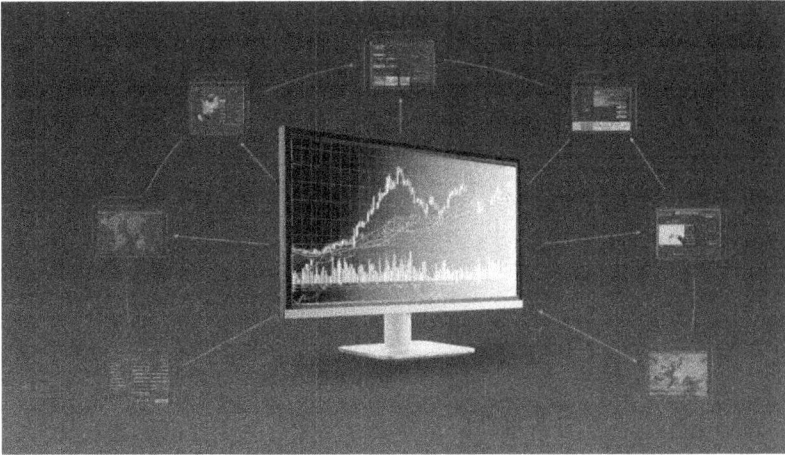

Reinforcement learning teaches trading agents to make decisions through trial and error. By continuously interacting with market data, these agents optimize their trading strategies over time to maximize returns while managing risk.

Market volatility presents significant challenges, but reinforcement learning can adapt through online learning — constantly updating models with new data. Adaptive exploration strategies help these models stay effective even in non-stationary, ever-changing market conditions.

Reinforcement learning isn't just theoretical — it's already being used in real-world applications. It optimizes portfolio management, improves algorithmic trade execution, and enables dynamic hedging strategies that adjust to market changes in real time.

Real-time sentiment analysis uses natural language processing to gauge investor emotions from text data. Machine learning models process this information instantly, helping traders anticipate market reactions.

Event-driven trading focuses on detecting key market-moving events — like earnings reports, geopolitical developments, or economic data releases. By acting quickly on these events, traders can take advantage of short-term price movements.

High-frequency trading faces several challenges. Low latency is crucial for timely signal generation. Market data noise must be filtered effectively. And maintaining data quality is essential for reliable trading signals. Despite these challenges, strong signal generation can unlock significant short-term market opportunities.

To conclude, advanced predictive analytics — including deep learning, large language models, multimodal fusion, and reinforcement learning — are transforming how we forecast and respond to market trends. By embracing these innovative techniques, traders and analysts can gain deeper insights, make faster decisions, and improve their overall market performance.

Chapter 7: AI in Portfolio Management

AI models play a key role in balancing risk and return. They dynamically assess financial data, allowing real-time adjustments to risk exposure. With predictive pattern recognition, AI spots market trends, which helps optimize how assets are allocated. This makes AI-driven asset allocation far more effective than traditional methods, enhancing portfolio performance.

AI uses machine learning techniques for predictive analytics. Regression analysis helps predict continuous outcomes like price trends. Classification methods forecast discrete events, such as whether an asset will outperform. Neural networks simulate the way the human brain processes information, boosting the accuracy of predictions in financial markets.

AI-driven optimization outperforms traditional methods by using sophisticated algorithms and deep data analysis. Unlike manual, historically focused portfolio management, AI can react dynamically. One of the biggest advantages is improved risk management — AI adjusts portfolios on the fly to help minimize downside risks and protect returns.

Robo-advisors have evolved from basic, rules-based systems to highly adaptive, AI-driven platforms. Early robo-advisors followed set algorithms. Today's versions use AI to tailor investment strategies based on real-time market data and individual investor profiles, offering a much more personalized approach.

AI enhances algorithmic trading in several ways. Self-learning strategies allow trading algorithms to adapt over time. AI also detects anomalies — unusual market patterns that might signal risk or opportunity. Plus, AI executes trades with greater speed and precision, giving traders a competitive edge in fast-moving markets.

AI enables true personalization by analyzing investor behavior and preferences. This lets platforms offer investment advice that's really tailored to each person. Plus, AI's ability to process real-time data supports decisions that are both timely and well-informed, improving outcomes for investors.

Feedback loops are a powerful AI tool in portfolio management. They allow AI systems to learn from portfolio performance and continually refine strategies. Through this ongoing learning, asset allocation becomes more effective over time. This dynamic adjustment helps maximize returns and keep the portfolio aligned with changing market conditions.

Real-time data is critical. AI models ingest a continuous stream of market information and economic indicators to inform decision-making. This allows for rapid portfolio adjustments, ensuring that investments remain aligned with shifting market dynamics and economic trends.

Performance monitoring and dynamic rebalancing are key strengths of AI systems. By constantly tracking performance, AI detects when a portfolio drifts from its target risk or return profile. Automated rebalancing then adjusts the portfolio, ensuring it stays aligned with investment goals — even as market conditions change.

Traditional derivatives pricing models, like Black-Scholes or binomial trees, rely on strict assumptions — such as constant volatility — which can limit their accuracy, especially for complex or exotic derivatives. These limitations have opened the door for AI-driven alternatives that better handle market complexity.

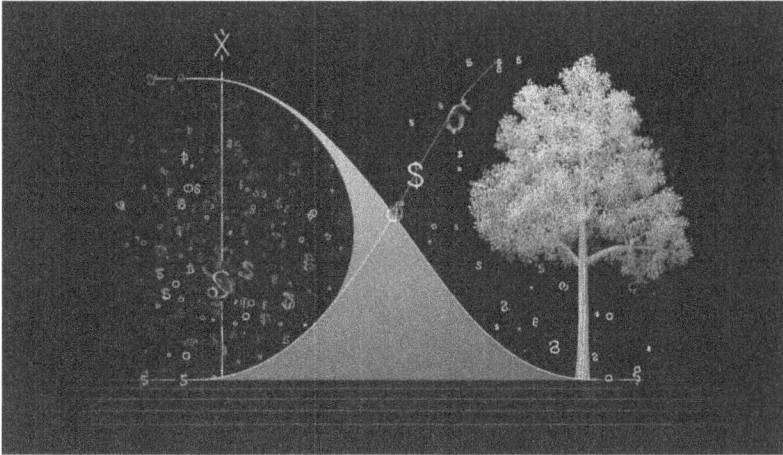

AI brings several advantages to derivatives pricing. It can handle nonlinear relationships in financial models, work with high-dimensional data sets, and adapt to complex market behaviors. This leads to more accurate and flexible pricing models for a wide range of derivative products.

In hedging, AI allows for dynamic adjustments based on real-time market data. This means hedge positions can be fine-tuned continuously, reducing risk exposure more effectively. Plus, AI-driven strategies optimize the cost of hedging, making it more efficient and adaptable to changing market conditions.

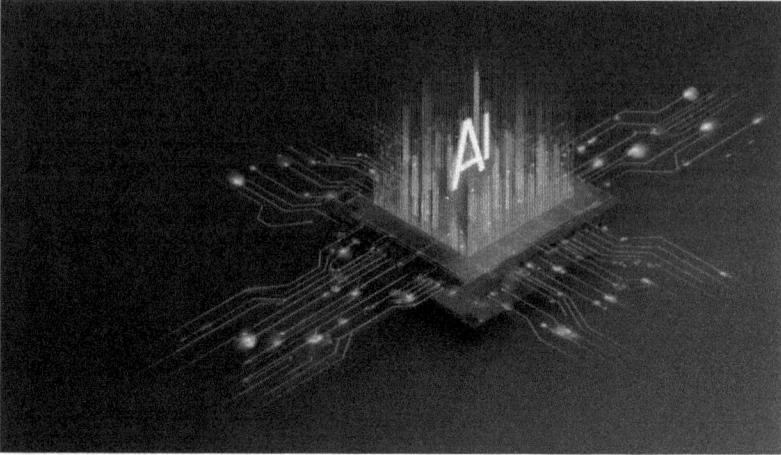

Multi-agent systems involve multiple autonomous agents working together to solve complex problems. In finance, they help model markets and support intricate decision-making processes. Through collaboration and communication, these agents can adapt strategies dynamically to shifting market environments.

Agents in a multi-agent system work cooperatively to mitigate risk. By sharing insights, they improve collective decision-making and help identify emerging market opportunities or threats. Coordinated action among agents strengthens a portfolio's ability to withstand market fluctuations.

In portfolio construction, multi-agent systems manage diversification more effectively. By coordinating independent agents, they ensure a broad range of assets and strategies are considered. This enhances both portfolio stability and performance, helping investors achieve better long-term outcomes.

Chapter 7: AI In Portfolio Management

To conclude, AI is transforming portfolio management in profound ways — from dynamic risk-return optimization and smarter robo-advisors, to adaptive asset allocation, advanced derivatives pricing, and multi-agent diversification strategies. These innovations are setting new standards for how portfolios are built, managed, and optimized in today's complex financial world.

Chapter 8: AI in ESG Analysis and Sustainable Investing

ESG stands for Environmental, Social, and Governance. These are three critical factors used to evaluate a company's sustainability and ethical impact. Environmental factors look at how a company affects nature — things like pollution, resource use, and climate change. Social factors consider relationships with employees, customers, and communities. Governance examines how a company is led — leadership quality, ethics, and transparency. Together, these provide a comprehensive view of corporate responsibility.

AI is transforming ESG analysis by tackling large and complex datasets that traditional methods struggle with. First, advanced data processing allows AI to sift through massive amounts of ESG information quickly, finding patterns that analysts might miss. Second, AI can uncover deeper insights, giving a richer, more accurate picture of a company's sustainability performance. And finally, by enhancing analysis, AI supports better decision-making, helping investors act on timely, comprehensive ESG intelligence.

Emerging trends in AI-driven sustainable investing are reshaping the field. AI automates the collection and analysis of ESG data, making processes faster and more accurate. It also boosts predictive analytics, enabling investors to forecast outcomes more reliably. And importantly, AI allows for more personalized investment products that align with individual sustainability preferences, which is a growing demand in today's market.

Traditional ESG scoring often relies on qualitative assessments, which can introduce subjectivity and bias. These methods also use fixed weightings for different factors, making it hard to adapt when market conditions or data change. As a result, traditional approaches can lack the flexibility and responsiveness needed in today's fast-paced environment.

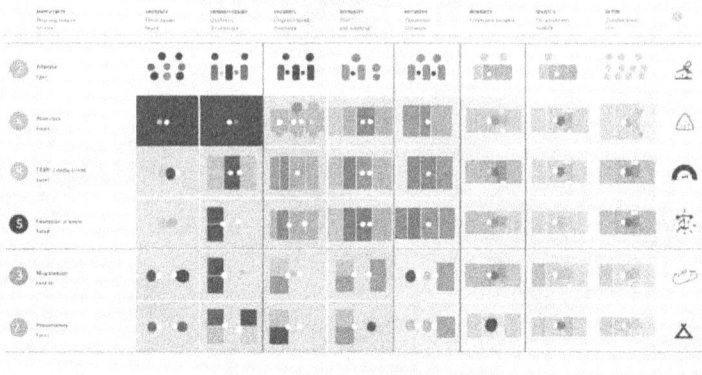

AI changes the game by using machine learning algorithms that can adjust factor weightings dynamically. As new data comes in, AI refines the scoring model to reflect real-time patterns. Over time, this means ESG scores become more accurate because AI continuously learns and fine-tunes its assessments, reducing the impact of outdated assumptions.

In practice, AI-driven ESG scoring systems have shown they can process complex data more effectively than traditional methods. Organizations using AI-powered scoring have reported better investment outcomes, thanks to improved insights. Real-world case studies demonstrate how companies across industries are applying AI tools to enhance their ESG evaluation and ultimately support more informed decision-making.

Natural Language Processing, or NLP, allows AI systems to handle vast amounts of text from ESG disclosures. It helps extract key information automatically, making sustainability assessments quicker and more precise. Additionally, sentiment analysis can evaluate the tone and language used in these disclosures, providing insights into the authenticity of a company's sustainability claims.

One of the key challenges in ESG investing is detecting greenwashing — when companies exaggerate or falsify their sustainability efforts. AI can spot inconsistencies between what companies report and what their actual performance shows. It also analyzes vague or ambiguous language in reports, helping to flag potential risks. This ensures that investments are based on truthful, reliable information.

Despite its benefits, NLP-based ESG analysis faces challenges. For instance, interpreting ambiguous language is still difficult for AI models, which can affect accuracy. Also, ESG reporting standards vary widely, making it hard to compare data consistently. Finally, the limited availability of labeled ESG data hinders the training of effective NLP models, meaning results may not always be reliable.

AI plays a crucial role in aligning investment portfolios with climate scenarios. Machine learning can analyze vast climate datasets to check whether investments are aligned with global goals like the Paris Agreement. AI models also simulate different emission scenarios, helping predict how investments might impact — or be affected by — future environmental outcomes, including temperature forecasts that guide sustainable investing decisions.

Forecasting carbon risks is another powerful application of AI. Machine learning can predict how carbon emissions will affect future regulatory environments and environmental risks. AI can also assess physical risks, like how extreme weather might impact investments. By anticipating these factors, AI helps investors identify potential financial risks and opportunities related to carbon exposure.

Integrating climate data into investment strategies means better risk management and smarter decisions. AI analyzes this data to offer actionable insights, helping investors avoid environmental risks. It also supports transitioning to low-carbon portfolios by guiding investments toward more sustainable assets, making climate-conscious investing more achievable.

AI helps investors align their portfolios with the United Nations Sustainable Development Goals, or SDGs. By analyzing investment data, AI tools can map where investments contribute to specific goals, like clean energy or reduced inequalities. This empowers investors to make choices that support global sustainability targets in a meaningful way.

For effective SDG tracking, AI relies on both structured data — such as company reports — and unstructured data from third-party sources. By integrating these data types, AI models provide a comprehensive picture of how investments are performing against sustainability indicators. This mixed approach ensures more robust and accurate SDG analysis.

AI tools also support the measurement and reporting of sustainable impact. By automating complex data analysis, AI delivers accurate impact assessments. This enhances transparency for investors and stakeholders, ensuring that sustainability claims are backed by credible data and helping build trust and accountability.

When it comes to assessing social impact, AI applies machine learning to analyze complex datasets on issues like labor practices and compliance. It can also evaluate community engagement, helping businesses improve their social strategies. Importantly, AI assesses adherence to human rights standards, ensuring that investments meet ethical expectations.

Integrating social metrics into investment decisions means AI helps investors include social impact data seamlessly into their portfolios. This supports investments that align with broader social objectives, such as community development or labor rights, ensuring that financial returns don't come at the cost of ethical considerations.

AI-powered social impact screening is already improving investment practices. It enhances the identification of responsible opportunities and supports better decision-making aligned with ethical and social values. By flagging potential risks related to social practices, AI also helps investors protect their reputations and reduce exposure to controversies.

Chapter 8: AI In ESG Analysis And Sustainable Investing

To conclude, AI is revolutionizing ESG analysis and sustainable investing by providing advanced tools for data processing, risk forecasting, and impact assessment. It addresses traditional challenges like subjectivity and data complexity, offering more accurate, efficient, and insightful analysis. As AI continues to evolve, it will empower investors to make smarter, more responsible decisions that contribute meaningfully to global sustainability goals.

Chapter 9: Hyper-Personalization Of Financial Services

AI enables granular customer segmentation by efficiently processing vast datasets to identify patterns in behavior and demographics. By combining behavioral, demographic, and transactional data, AI can create detailed customer segments. This means financial services can be tailored much more precisely to meet the unique needs of each customer.

AI doesn't just segment customers—it actively matches financial products to their needs. By analyzing client data, AI recommends suitable products dynamically and adapts those recommendations as client preferences change. This continuous assessment ensures that customers always see the most relevant products for their goals and risk tolerance.

While dynamic matching offers huge benefits like better engagement and conversion rates, it comes with challenges. Data quality must be consistently high, which isn't always easy. Integrating AI systems with existing platforms can also be complex, and privacy and compliance must be carefully managed to protect customer data.

Real-time recommendation engines analyze live data streams to understand customer behaviors instantly. They deliver immediate product suggestions tailored to the customer's preferences and actions in that moment. This real-time approach enhances the customer experience by making recommendations that feel timely and relevant.

Several key data sources power financial product recommendations. Transactional data shows spending patterns. Customer interactions provide insights into behavior. Social media can reveal preferences and emerging trends. Market data helps ensure recommendations remain aligned with broader economic changes—all combining for highly tailored suggestions.

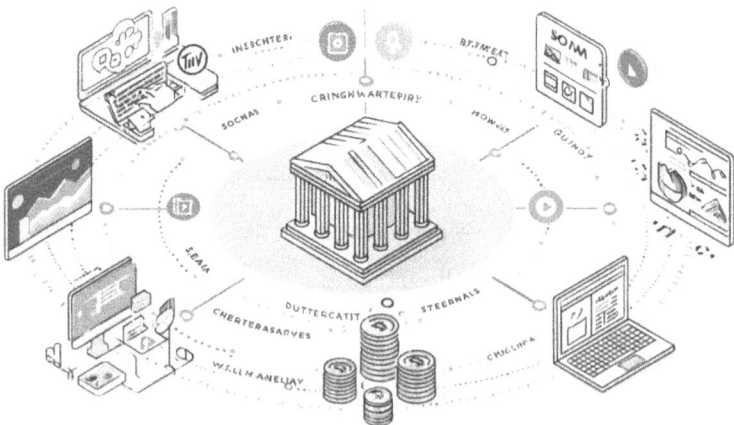

Chapter 9: Hyper-Personalisation of Financial Services

Real-time recommendations have a direct impact on customer engagement. They make experiences feel more personal, encouraging customers to interact more often and more meaningfully. This, in turn, drives higher conversion rates, as recommendations better meet customer needs at the right time.

Understanding customer behavior is essential in finance. AI analyses patterns of user behavior, making it possible to predict future actions and preferences. This allows financial institutions to offer proactive advice and recommendations that meet customers' evolving needs.

Intent modelling allows AI to anticipate customer needs by predicting likely future actions. This means financial products can be suggested at just the right moment, increasing the chances that the customer will find them helpful and relevant to their situation.

Predictive analytics is already making a difference in finance. It helps with targeting valuable customers, optimizing risk assessment, and offering personalized financial planning. These case studies show how analytics not only improves business outcomes but also enhances the customer experience.

Micro-moments are those brief, critical points when customers are actively seeking information or making decisions. These are powerful opportunities for financial service providers to engage customers with timely, relevant offers or support—right when they're most receptive.

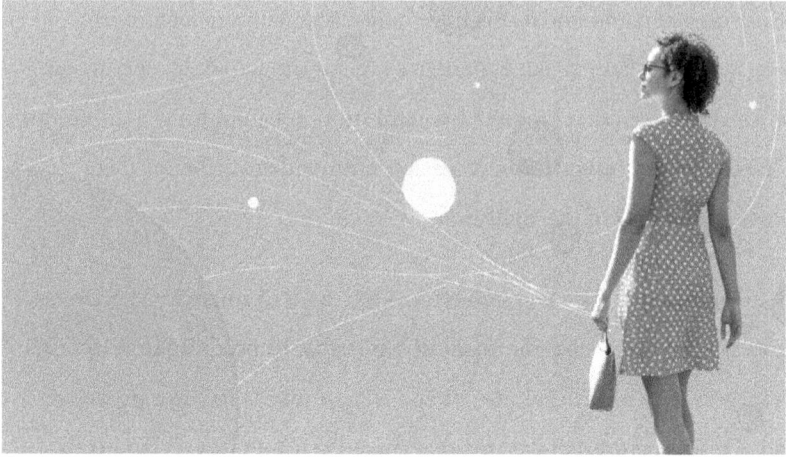

AI-driven lifecycle marketing strategies enable financial institutions to engage customers throughout their journey. From acquisition—by predicting potential interests—to engagement and retention, AI helps deliver the right message at the right time, keeping customers engaged and loyal.

Contextual targeting uses real-time data to understand a customer's current environment and preferences. This allows for highly relevant messaging that resonates with customers. When done right, it significantly boosts marketing effectiveness by aligning offers with the customer's immediate needs.

Responsible nudging in finance means encouraging positive behaviors without being manipulative. It's about offering helpful suggestions while respecting customer choice and maintaining transparency. Done ethically, nudging builds trust and supports long-term relationships.

Over-personalization carries risks. It can lead to privacy invasions, bias in recommendations, and discomfort for customers who feel their data is overused. Transparency, fairness, and respect for customer autonomy are essential to prevent these risks and maintain trust.

Regulatory compliance and ethical standards are vital when using personalization in finance. Following regulations protects customers and ensures transparency. Ethical practices also help maintain trust and safeguard customer relationships in the long term.

To conclude, AI-driven personalization has the power to transform financial services—enhancing engagement, satisfaction, and business results. But it must be handled responsibly, balancing innovation with privacy, compliance, and ethics to truly benefit both providers and customers.

Chapter 10: Conversational AI and Virtual Advisors

Let's start by defining traditional chatbots. These operate on predefined rules, following scripts to handle customer queries. They're limited to basic tasks and lack the ability to understand deeper conversation context. This means their responses often feel robotic and can't adapt beyond their programmed scenarios.

Now, generative virtual financial advisors bring a significant leap forward. These AI-driven systems offer personalized financial advice, understanding individual customer goals and profiles. They can interpret complex queries and hold multi-turn, natural conversations, adjusting their responses as the conversation evolves. This makes interactions feel more human and helpful.

When we compare the two, generative advisors clearly offer a richer experience. They improve customer satisfaction by providing tailored recommendations and more engaging conversations. On the other hand, traditional chatbots still have a role in quickly handling routine, transactional requests, where speed and efficiency are key.

Multilingual AI agents play a crucial role in improving customer engagement, especially in diverse markets. By supporting multiple languages, they allow banks to connect with customers in their preferred language, fostering stronger relationships and trust.

These AI agents can be deployed across both digital and physical banking channels. This means customers experience consistent support whether they're on a website, using a mobile app, or visiting a branch. Plus, AI-powered personalization ensures each interaction feels tailored to the individual.

However, there are challenges. AI must understand language nuances and context to communicate effectively. Seamless integration across platforms is essential, which requires robust backend systems. Advanced natural language processing and continuous learning help overcome these barriers, ensuring better service automation and customer satisfaction.

Voice biometrics rely on the unique characteristics of each person's voice for secure identity verification. This technology allows hands-free authentication, making the process both convenient and secure. It's highly accurate and requires minimal effort from users, making it an ideal solution for customer verification.

In banking, voice biometrics can be used for verifying customers during call center interactions, allowing secure access to services. It also supports secure login for mobile apps and plays a key role in preventing fraud by ensuring only legitimate users can perform transactions.

Security and privacy are top concerns. Voice data must be protected with robust security measures. Banks need to be transparent with customers about how their voice data is used and ensure compliance with all regulatory standards. This builds trust and helps avoid legal issues.

In wealth advisory, AI provides personalized investment recommendations based on a client's goals and financial situation. This allows for highly tailored advice that can adapt as the client's needs change over time.

AI also streamlines retirement planning by automating tasks like portfolio rebalancing and goal tracking. It ensures portfolios remain aligned with market conditions and customer objectives, making long-term financial planning more efficient and responsive.

While AI-driven wealth advisory offers scalable, data-driven advice to a broad client base, there are limitations. AI's decision-making processes can be hard to interpret, and human oversight remains essential. Additionally, regulatory compliance is a constant challenge, requiring careful management of AI advisory services.

Emotion detection technology uses AI to analyze both speech patterns and text, identifying sentiment and emotional states. This allows systems to better understand customer feelings, which is essential for delivering empathetic responses and building stronger customer relationships.

By integrating emotion recognition, AI agents can adjust their responses to match the customer's emotional state. This helps provide a more personalized experience and ensures sensitive issues are escalated to human agents when necessary, improving overall service quality.

Emotion-aware systems enhance customer loyalty by making interactions feel more genuine and human. They also help resolve problems more effectively, as empathetic communication can defuse tense situations and foster positive outcomes.

To wrap up, conversational AI and virtual advisors are reshaping financial services. They improve customer engagement, enhance security, personalize advisory services, and even bring empathy into digital interactions. As these technologies evolve, they'll continue to redefine how financial institutions serve and support their customers.

Chapter 11: Behavioral Analytics and Customer Retention

Behavioral analytics plays a crucial role in customer retention by allowing financial organizations to gather and analyze customer activity data. By recognizing patterns in customer financial behavior, we can anticipate their needs and provide services that build loyalty. This proactive approach strengthens relationships and encourages long-term engagement.

Data-driven insights are key to retaining customers. By identifying customers who are at risk of leaving through analyzing their behaviors and engagement patterns, organizations can take targeted action. These personalized retention strategies help maintain customer loyalty, especially when combined with timely, proactive engagement based on data insights.

When behavioral analytics is integrated with Customer Relationship Management systems, we can significantly enhance customer interactions. Personalized engagements make customers feel valued, while automated journey mapping helps organizations track and manage the customer experience. This integration leads to improved service delivery and strengthens both retention and engagement efforts.

AI allows us to analyze complex, multi-dimensional customer data effectively. By harnessing AI, we can build a deeper understanding of customer behavior, enabling highly personalized engagement that resonates with individual customers.

Machine learning plays a vital role in personalization by detecting subtle patterns in customer behavior. This allows us to make tailored recommendations that increase relevance and customer satisfaction. As a result, personalized communication boosts engagement and helps foster stronger connections with customers.

Dynamic customer profiles, powered by real-time data updates, give marketers the flexibility to adapt their strategies quickly. By continuously refining customer insights, we can ensure that engagement remains relevant and personalized, leading to better customer experiences and stronger loyalty.

Churn prediction starts with identifying key indicators like declining engagement, reduced transaction frequency, and shifts in customer sentiment. By closely monitoring these factors, organizations can flag customers at risk of leaving and act before it's too late.

Predictive models analyze both historical and current data to accurately assess a customer's likelihood of churning. These models allow businesses to assign churn risk scores, helping them focus their retention efforts on high-risk customers and prioritize outreach accordingly.

To reduce churn effectively, companies need to implement experience optimization strategies. This includes offering personalized deals, improving service accessibility across various channels, and maintaining timely, proactive communication with customers. These actions make customers feel valued and engaged, reducing the risk of churn.

Behavioral analytics can even detect major life events—like marriage or childbirth—by analyzing patterns in customer behavior. These life events often impact financial needs, presenting opportunities for organizations to offer tailored products and services that meet changing customer requirements.

When we tailor financial products and advice based on detected life events, we can provide highly relevant solutions that meet customers' needs at critical moments. This not only enhances customer satisfaction but also strengthens long-term relationships and loyalty.

Proactive outreach during key life events helps build lasting customer loyalty. By anticipating needs and positioning themselves as trusted advisors, organizations foster stronger relationships and become a go-to resource for customers over time.

AI is powerful in mining unstructured survey responses and feedback data. It can efficiently extract themes and patterns, and perform sentiment analysis to understand both positive and negative customer feelings. These insights are invaluable for making informed business decisions.

Sentiment analysis enables organizations to evaluate customer emotions in real-time, providing immediate insights into satisfaction levels and potential concerns. This dynamic understanding supports more responsive and effective customer engagement.

Closing the feedback loop means acting on the insights gained from analysis. By transforming customer feedback into actionable improvements, organizations can enhance products, services, and customer experiences—showing customers that their opinions truly matter.

AI-driven micro-segmentation allows us to create highly specific customer groups based on shared characteristics. This granularity enables marketers to deliver highly targeted campaigns that resonate deeply with each group, increasing engagement and conversion rates.

Predictive analytics helps forecast customer behavior, enabling organizations to tailor their marketing strategies accordingly. By delivering offers that are personalized and predictive, businesses can boost customer engagement, increase conversion, and build lasting loyalty.

To ensure campaigns remain effective, continuous measurement is essential. By analyzing performance data, marketers can refine their strategies over time, ensuring ongoing improvement and maximizing return on investment.

Chapter 11: Behavioural Analytics and Customer Retention

In conclusion, leveraging behavioral analytics and AI empowers organizations to enhance customer retention and engagement through deeper insights, proactive strategies, and personalized customer experiences. By applying these data-driven approaches, businesses can build stronger relationships and secure long-term loyalty in an increasingly competitive market.

Chapter 12: Bias, Fairness, and Responsible AI in Finance

Bias in financial models can come from several common sources. Unrepresentative data is a big issue—if the data used to train a model doesn't reflect the diversity of real-world populations or scenarios, the model can make biased decisions. Historical prejudices can also sneak into the data, carrying forward discriminatory practices from the past. And finally, bias can arise from algorithmic design choices—sometimes, the way an algorithm is set up can unintentionally reinforce unfairness.

There are several methods we can use to detect bias in financial models. Fairness metrics help us quantitatively assess if a model is treating groups equitably. Bias testing frameworks provide structured ways to check models for bias during evaluation. And adversarial testing puts models through challenging scenarios to expose hidden biases that might not show up in regular testing.

Once we've detected bias, there are various strategies we can apply to mitigate it. Data balancing techniques help ensure different groups are fairly represented in the data. We can adjust algorithms to correct biased behavior and promote fairness. Enhancing transparency helps users understand how decisions are made, which builds trust. And finally, continuous monitoring lets us catch and address new biases as they emerge over time.

When it comes to model auditing in finance, there are several key techniques. Model validation checks that models are delivering accurate and reliable results across different situations. Performance evaluation helps us spot weaknesses and improve predictive accuracy. And compliance checks make sure our models meet both regulatory and ethical standards.

Drift detection is all about maintaining the integrity of financial models over time. It helps us spot changes in data patterns or model behavior before they become a problem. This way, we can make timely updates to the model to avoid any drop in performance or fairness.

Effective governance is essential for responsible AI in finance. Governance frameworks define the policies that guide ethical AI development. They also set out clear roles and responsibilities for everyone involved in the AI lifecycle—from development through to deployment and maintenance. Governance ensures that there are processes in place for continuous monitoring and auditing, keeping AI models compliant and effective.

Model explainability is critical for all stakeholders in finance. Regulators need transparency to assess models for fairness and compliance. Clients benefit when they can understand model outputs, which builds confidence and trust. And within internal teams, explainability helps improve collaboration, model development, and informed decision-making.

In credit scoring, it's important to use techniques that enhance transparency. Clear documentation explains how models work and why they make certain decisions. Using accessible explanations—free from technical jargon—helps diverse audiences understand model outputs. Finally, maintaining open dialogue with stakeholders allows for feedback, addresses concerns, and reinforces fairness.

Ethics committees play a key role in AI governance. They provide oversight by reviewing AI projects and ensuring responsible development. They make sure projects align with established ethical standards to minimize risks. And they help identify and address potential risks before they become issues during AI deployment.

Model registries bring several important benefits. They serve as a centralized place to store all model documentation, making information easier to access and manage. They also support version control, which helps track updates and changes over time. And they play a role in oversight by supporting compliance and ensuring that models maintain their integrity.

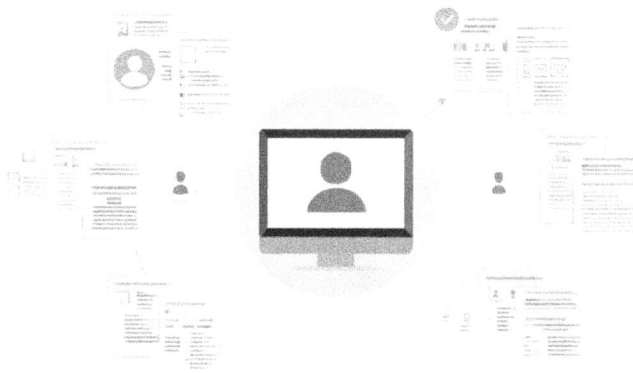

Audit trails are essential for accountability in AI systems. They document every step of model development, providing transparency and traceability. Audit trails also track deployments and updates, which helps maintain accountability throughout the model's lifecycle. They're invaluable during investigations, providing detailed records of AI activities. Overall, maintaining strong audit trails shows a commitment to responsible and ethical AI practices.

The AI Act sets out legal requirements for AI use in the financial sector, especially for high-risk systems. It mandates transparency so stakeholders and regulators can understand how models work. Risk management is another key area—it requires comprehensive assessments and mitigation strategies. And importantly, human oversight must be part of AI systems to prevent mistakes and ensure ethical decision-making.

ISO/IEC 42001 provides guidelines for managing AI systems responsibly. This standard helps organizations establish systems that promote ethical behavior, transparency, and accountability in AI. By following these guidelines, financial institutions can ensure their AI systems meet expectations for responsible operation.

Beyond global standards, it's crucial to comply with local and regional regulations. These may have specific requirements for AI in finance, and understanding them is essential for maintaining compliance and avoiding legal issues.

To wrap up, responsible AI in finance isn't just a technical challenge—it's an ongoing commitment to fairness, transparency, and ethical governance. By identifying and mitigating bias, conducting regular audits, building explainability into models, and aligning with both global standards and local regulations, financial institutions can foster trust and integrity in their AI systems.

Chapter 13: AI-Augmented Roles in Financial Institutions

Data analysts are moving away from manual data processing. Instead of spending time on repetitive tasks, they're now focusing on higher-level analysis and interpreting insights generated by AI. This shift requires them to develop stronger technical and analytical skills, particularly in working alongside AI systems to make better business decisions.

With AI transforming traditional roles, we're seeing the rise of the AI strategist — a role focused on leveraging AI for business strategy. These professionals help organizations navigate AI adoption, align AI initiatives with business goals, and manage the integration of AI into daily operations.

Chapter 13: AI-Augmented Roles in Financial Institutions

Financial professionals of the future will need a mix of key skills. They'll need a solid understanding of AI concepts, strong data literacy to interpret complex datasets, critical thinking to make sound judgments, and adaptability to stay relevant as technology and financial landscapes evolve rapidly.

Prompt engineering is becoming an essential skill in finance. It involves crafting effective inputs or prompts that guide AI models to generate accurate and useful outputs. In financial contexts, this ensures that AI tools provide reliable analysis and support informed decision-making.

AI agents don't manage themselves — they require human oversight. This includes maintaining the quality of AI outputs, preventing errors, and ensuring that AI-driven processes comply with regulations. Oversight plays a critical role in minimizing risks and making sure AI works as intended in financial workflows.

Transparency and fairness are key when using AI in finance. Professionals need to ensure that AI outputs are understandable, explainable, and free from bias. This helps build trust with clients and regulators and reduces the risk of unintended consequences from AI decisions.

AI copilot tools are transforming how financial professionals work. These tools automate routine tasks, freeing up time for more complex work. They also generate insights by analyzing large amounts of data and offer decision support by providing recommendations and predictive analytics — all of which enhance daily operations.

For advisors, AI copilots are enhancing client engagement by providing timely insights and support. They help deliver personalized advisory services, strengthen client relationships, and add value through AI-assisted interactions and advice.

Auditors and compliance officers also benefit from AI copilots. These tools speed up data analysis, help detect anomalies or fraud, and automate compliance checks. This reduces manual workload, improves accuracy, and helps ensure that institutions meet regulatory requirements.

AI is also revolutionizing staff training through personalized learning. By tailoring training to each employee's needs, AI ensures better engagement and more effective skill development. It helps identify knowledge gaps and delivers customized content, making learning more efficient and targeted.

Adaptive learning platforms go a step further by adjusting content in real-time based on an individual's progress. This dynamic approach keeps employees engaged and ensures that training meets their specific learning needs, ultimately improving learning outcomes within financial institutions.

It's important to measure the effectiveness of training. AI helps track learning progress, identifies areas needing improvement, and assesses the return on investment in training programs. This ensures that skill development efforts are impactful and aligned with business goals.

Identifying skills gaps is the first step in effective reskilling. By understanding what skills are missing now — and what will be needed in the future — organizations can design targeted reskilling programs. This helps employees transition into new, AI-augmented roles with confidence.

As AI reshapes workflows, new job roles must be designed to complement AI systems rather than compete with them. This means rethinking job responsibilities and ensuring that human skills are used where they add the most value alongside AI capabilities.

Looking at real-world examples, we see how successful organizations integrate AI by planning strategically, investing in training, and adopting new technologies. These case studies show both the opportunities and the challenges of workforce transformation in the AI era — offering valuable lessons for others on a similar journey.

To wrap up, AI is fundamentally changing roles in financial institutions. By evolving skill sets, leveraging new tools, and continuously learning, professionals and organizations can thrive in this new landscape. Embracing AI as a partner — not a replacement — is key to staying competitive and delivering value in the age of automation.

Chapter 14: Strategic AI Adoption and Digital Transformation

When it comes to AI operating models, financial institutions generally choose between centralized and decentralized governance. Centralized AI governance ensures consistency, strict control, and unified policies across the organization — this is especially important for regulatory compliance and risk management. On the other hand, decentralized governance gives business units flexibility to tailor AI solutions for their specific needs, promoting agility and faster innovation at the local level.

Integrating AI into existing business processes is key to maximizing its value. First, AI boosts operational efficiency by automating repetitive tasks and reducing errors. Second, it enhances decision-making by analyzing data and providing predictive insights. And finally, when AI is seamlessly aligned with business goals, it delivers more meaningful and measurable value to the organization.

Successful AI adoption requires not just technology but also talent and collaboration. Attracting skilled AI professionals is critical — their expertise drives innovation and competitive advantage. But it's equally important to foster cross-functional collaboration between data scientists, business leaders, and IT teams. This collaboration helps ensure AI projects meet business needs and are effectively implemented.

Building an AI roadmap starts with identifying high-impact use cases. Focus on AI applications that drive revenue growth, such as those that open new markets or improve customer offerings. Select use cases that reduce risks, like fraud detection or regulatory compliance. And prioritize solutions that improve the customer experience to boost satisfaction and long-term loyalty.

Defining clear milestones and KPIs is crucial for AI initiatives. Milestones create structured checkpoints to track progress and help ensure timely delivery. KPIs help measure success and ensure AI projects align with strategic goals. Tracking both helps manage resources effectively and mitigates risks during project execution.

To justify AI investments, you need clear ROI models. Start with quantitative measures — financial metrics that show how AI improves business performance. But don't overlook qualitative benefits, like better customer satisfaction and operational efficiencies. When communicating ROI, present insights clearly to stakeholders to secure their ongoing support and future investment.

Innovation labs serve as dedicated spaces for AI exploration and experimentation. They provide an environment for testing new technologies and validating concepts before scaling. By streamlining processes, these labs can significantly shorten the time it takes to bring new AI solutions to market.

A key purpose of innovation labs is to foster a culture of experimentation. They accelerate AI project prototyping, allowing teams to test ideas quickly and refine them based on feedback. And once AI pilots prove successful, labs help organizations scale those solutions effectively across business units to maximize impact.

When partnering with FinTechs and AI vendors, it's important to choose wisely. First, ensure the technology fits your AI needs and existing infrastructure. Evaluate the partner's expertise and domain knowledge to make sure they can deliver quality results. Always verify that they meet legal and compliance standards. And finally, make sure the partner's strategic vision aligns with your organization's goals.

Establishing effective collaboration models with external partners is critical. Partnerships should be structured to promote clear communication, shared goals, and joint success. This includes defining responsibilities, workflows, and governance structures up front to avoid misunderstandings and maximize value.

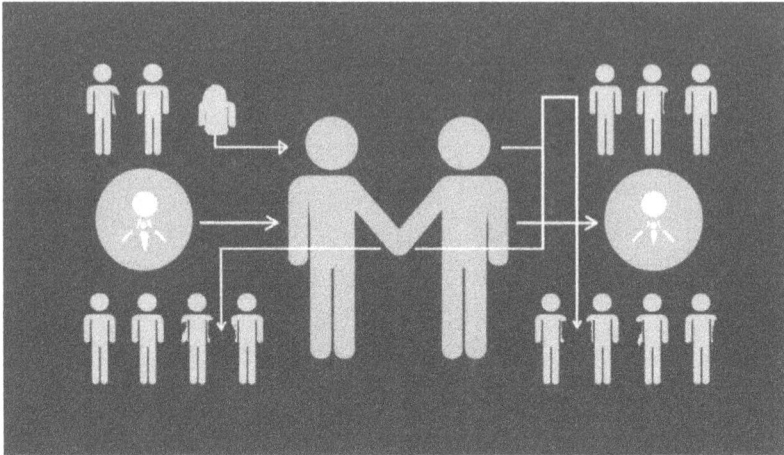

Managing risk and compliance is a key part of any AI partnership. You must ensure the partner adheres to all relevant regulations — this protects both your organization and your customers. Operational risks also need to be identified and mitigated early to prevent disruptions and ensure business continuity.

AI capability maturity models help organizations measure and improve their AI readiness. They assess governance to ensure ethical practices and oversight. They evaluate technology infrastructure to confirm it supports scalable AI systems. They look at talent — ensuring the right skills are in place. They examine data management, which is critical for AI success. And they measure the deployment of AI use cases to evaluate real-world impact.

Benchmarking AI capabilities provides valuable insights for continuous improvement. It allows you to compare your organization's AI maturity against industry standards and peers. This helps in setting realistic adoption targets and guides strategic implementation of AI across the business.

Continuous improvement is essential for scaling AI effectively. Regular assessments help align AI initiatives with changing business goals and technology trends. The refinement process ensures your AI systems remain optimized and relevant. And adopting best practices for scaling allows your AI solutions to grow alongside your business, maximizing their long-term value.

To wrap up, strategic AI adoption in financial services requires a clear roadmap, the right operating model, strong partnerships, and ongoing capability assessments. With the right approach, organizations can unlock AI's full potential — driving innovation, enhancing customer experience, reducing risks, and achieving sustainable growth. Thank you for joining me in this session, and I look forward to seeing you in the next module!

www.ingramcontent.com/pod-product-compliance
Lightning Source LLC
Chambersburg PA
CBHW060630210326
41520CB00010B/1542